PENGUIN CANADA

75 WAYS TO SAVE GAS

JIM DAVIDSON is president and founder of Car$mart Inc., a Canadian car brokerage and consulting firm. Jim has been helping frustrated car buyers for almost 20 years. He has franchised his novel brokerage concept all over North America, has built and sold a dot-com internet service, and now has become a specialist in "green" cars. Jim is frequently asked to comment on the auto industry on TV, radio, and in print. He lives in Toronto with his wife, Jennifer, and two children, Emily and Taylor.

75 WAYS to SAVE GAS

Clean, Green Tips to Cut Your Fuel Bill

Jim Davidson

PENGUIN
CANADA

PENGUIN CANADA

Published by the Penguin Group

Penguin Group (Canada), 90 Eglinton Avenue East, Suite 700,
Toronto, Ontario, Canada M4P 2Y3 (a division of Pearson Canada Inc.)

Penguin Group (USA) Inc., 375 Hudson Street, New York,
New York 10014, U.S.A.
Penguin Books Ltd, 80 Strand, London WC2R 0RL, England
Penguin Ireland, 25 St Stephen's Green, Dublin 2, Ireland
(a division of Penguin Books Ltd)
Penguin Group (Australia), 250 Camberwell Road, Camberwell, Victoria 3124,
Australia (a division of Pearson Australia Group Pty Ltd)
Penguin Books India Pvt Ltd, 11 Community Centre, Panchsheel Park,
New Delhi – 110 017, India
Penguin Group (NZ), 67 Apollo Drive, Rosedale, North Shore 0745,
Auckland, New Zealand (a division of Pearson New Zealand Ltd)
Penguin Books (South Africa) (Pty) Ltd, 24 Sturdee Avenue, Rosebank,
Johannesburg 2196, South Africa

Penguin Books Ltd, Registered Offices:
80 Strand, London WC2R 0RL, England

First published 2009

1 2 3 4 5 6 7 8 9 10

Manufactured in the U.S.A.

ISBN: 978-0-14-317205-5

Library and Archives Canada Cataloguing in Publication data available upon
request to the publisher.

Visit the Penguin Group (Canada) website at **www.penguin.ca**

Special and corporate bulk purchase rates available; please see
www.penguin.ca/corporatesales or call 1-800-810-3104, ext. 477 or 474

This book is dedicated to people everywhere who are sick and tired of paying high gas prices.

Fight back and take control of the fuel you use.

Enjoy the savings!

CONTENTS

75 WAYS TO SAVE GAS

Introduction

Gas prices are extremely volatile. We have seen the price of fuel reach and surpass record levels recently, whether because of economic uncertainty, political upheaval, or natural disasters. Some days, the price of fuel fluctuates radically, but overall, the price of gas is continually going up, and at an alarming rate. The times when gas drops in price are usually short-lived; higher gas prices in general are here to stay, so we need to look for ways in which to save and keep our hard-earned money in our pockets.

To many of us, the car is vital to our daily lives. But, next to our home, it is the biggest drain on our budgets. When it comes to our cars, it often feels like we can't live with them and can't live without them. So what can we do?

With more than 20 years of experience helping consumers make wise choices when it comes to automobiles, I've learned many cost-saving tips and tricks. *75 Ways to Save Gas* is chock full of easy-to-follow suggestions to help you save fuel—and potentially hundreds, if not thousands, of dollars each year on your gas bill. The book is divided into two parts. The first half focuses on fuel-saving tips that you can use before you even get into your car. The second half offers fuel-efficient ideas that you can take with you onto the road. At the back of the book, you'll find a list of helpful websites that you can visit for more ideas on how to further your savings.

My tips are easy to incorporate into your daily driving routine. You'll be surprised at the drop in your gas bill when you adopt even just a few of these fuel-saving methods. You'll experience instant financial savings. So take control of the high cost of fuel and start saving money today.

PART 1

Alternatives to
using your car

1 Why not walk?

For your local errands and activities, why not use your feet instead of your car to get things done? If you give yourself a bit of extra time and plan ahead, you'll be surprised how much you can get done by walking. Don't pollute. Instead, leave the car at home and save a lot of gas.

"Approximately 50% of car use is for trips within 3 miles of the home."
www.eartheasy.com

Here are seven ways to include walking in your busy schedule:

- Walk your kids one way to school every other day.
- Every second weekend, do your errands on foot.
- On your lunch break, walk from work to do errands.
- Walk with your partner or a friend to the local coffee shop.
- Walk to the corner store or video store instead of driving.
- Walk to and from the gym—it's also a great way to warm up and cool down.
- If you live close enough, walk to work at least once a week.

In addition to saving fuel, you'll enjoy the fresh air, peace of mind, and health benefits that come with exercise.

TWELVE WALKING SAFETY TIPS

1. Walk facing traffic.
2. Use the sidewalks.
3. Don't jaywalk; always cross at intersections.
4. At night, walk on well-lit streets or paths.
5. Walk with a buddy.
6. Always be aware of your surroundings.
7. Carry a fully charged cell phone in case of emergencies.
8. Turn your mp3 player down or off so that you're able to hear what's going on.
9. Plan your route with safety in mind and avoid hazardous short-cuts.
10. Dress for the weather—make sure to layer up in winter and wear breathable fabrics in summer.
11. Wear comfortable shoes.
12. If you're going for a long walk, stay hydrated by taking some water with you.

2 Get on your bike

Most of us own or have access to a bicycle but don't use this gas- and pollution-free mode of transportation for our daily grind. Bikes are a great alternative to the car for commuting, shopping, and running errands. The savings from incorporating biking into your lifestyle can be staggering.

> "You save $15.80 per week
> by biking to work."
>
> www.biketowork.ca

This savings is based just on the cost of gas (in 2007). Add on the cost of your monthly car payments, insurance premium, and parking, and the price of owning a car goes up exponentially. Biking is becoming a very popular way to get around in urban areas. Most major cities in North America are committed to this new movement and are adding more and more kilometres of bike lanes to the roads each year. Bicyclists are slowly gaining more respect from motor vehicle drivers, and bike lanes are helping to make it easy and safe to get around on two wheels instead of four.

SAFE BIKING TIPS

BEFORE YOU RIDE: Tips from the League
of American Bicyclists

1. **Bike size and type.** Make sure your bike fits you. A bike frame
   that is too small or too big for you can be dangerous. Also make
   sure that the type of bike you ride is suitable to the terrain and
   your style of riding. A racing bike is not a good idea if you take
   back roads and trails to work!
2. **Helmet.** This is obvious—and the law in many provinces. Make
   sure your helmet is the right size for you; a comfortable, snug fit
   is essential.
3. **Lights and reflectors.** Even if you don't plan to ride at night, be
   prepared; having front and back lights and as many reflectors as
   possible on your bike may save your life. Be smart: Be seen.
4. **Bell.** Is a bell necessary on a bicycle? In most provinces it is. Make
   sure you have one. It's a great warning device for other bicyclists,
   vehicle drivers, and animals.
5. **Bright clothing.** Regardless of what time of day you ride your
   bike, bright clothing or reflective items will help you be seen by
   others. A reflective vest worn overtop your clothing is also a good
   idea.
6. **ABC quick check.** Each time before you ride, inspect your bike
   to make sure it's safe. "A" stands for check the *air* in your tires,
   "B" stands for check your *brakes*, and "C" stands for check your
   crank and *chain*. The "quick" part means inspect your quick-
   release wheel fasteners. You want to make sure that these are on
   correctly and tight.

WHEN RIDING: Tips from the League of American Bicyclists

1. **Always ride in the same direction as traffic.** Riding in the opposite direction is just asking for an accident.
2. **Obey all road signs.** Bicyclists must obey all the traffic signage and signals and road markings, just like drivers of motor vehicles.
3. **Ride defensively.** Always be prepared. Don't assume that everyone else on the road sees you. Ride ready for danger and be extra cautious at intersections.
4. **Know the law.** As a bicyclist, it's your responsibility to know the law as it pertains to bicyclists. In most provinces, bikes are considered to be the same as cars. Don't assume that you can glide past a stop sign or ride on the sidewalk—know what is legal and illegal.
5. **Avoid road hazards.** Always be on the lookout for potholes, sewer covers, speed bumps, and debris. It may be okay to hit this stuff in a car, but not on a bike with skinny tires!
6. **Be predictable.** Ride in a straight line, don't weave, and don't make erratic manoeuvres. Signal well in advance of your moves. Establish eye contact with drivers. One miscalculated move may mean the end of your bicycling career.

Need advice on bike gear? Want information on bike lanes and trails in your area? Check out these helpful sites:

British Columbia	www.cyclingbc.net	Ontario	www.ontariocycling.org
Yukon	www.sportyukon.com/ sportlinks	Quebec	www.fqsc.net/accueil.htm
Northwest Territories	www.canadatrails.ca/ biking/bike_nw.html	New Brunswick	www.velo.nb.ca
Alberta	www.albertabicycle.ab.ca	Newfoundland Labrador	www.bnl.nf.ca
Sasktachewan	www.saskcycling.ca	Nova Scotia	www.bicycle.ns.ca
Manitoba	www.cycling.mb.ca	Prince Edward Island	www.cpei.ca

3 Use transit

Public transit is cheap, relatively efficient, and by far the
most stress-free way to travel in big urban areas. You can
use the travel time to read, listen to music, organize your
work day, answer emails, relax, or just catch up on your
thoughts. Yes, transit can be crowded and slow, but it's
much more economical than owning a car and paying for
fuel.

> "A person can save more than $8,000
> per year by taking public transportation
> instead of driving."
>
> "Transit News" press release, American Public Transportation
> Association, July 31, 2008

On most transit systems you can now bring your bike,
to combine the benefits of these two gas-saving modes of
transportation. If using transit doesn't work for you on a
regular basis, try to use it on the weekends for some of
your errands and activities. You'll be surprised at how
easy it is to make the change—and make a difference.

> "A single city bus can take 40 vehicles
> off of the road, save 10,646 litres of fuel
> and keep 25 tonnes of greenhouse gases
> out of the atmosphere each year."
>
> The Auto$mart Guide, Natural Resources Canada, 2007

4 Take a taxi

Yes, riding in cabs can be expensive. But in many cases it's much cheaper than what you would spend on monthly car payments, gas, parking, and insurance if you owned a car. Many urban dwellers don't own cars. Instead, they get around the city in taxis and still save a lot of money. For example, a cab from West Vancouver to downtown Vancouver costs about $11.20; the average daily parking price in downtown Vancouver is $13.25.

Many of the taxis on the road today are fuel efficient and have low emissions, running on propane, natural gas, diesel, or hybrid power. Most taxi companies, understanding that bicyclists need a lift now and then, allow you to load your bike in the cab. So you can bike downhill to your destination and then hop in a cab with your bike for the uphill trek back home. To cut the cost in half, share the cab with a friend.

Different cab companies have different rules and regulations, but here are some things that you should expect from every cab ride:

- The driver should comply with all rules of the road.
- The interior of the cab should be smoke-free.
- If requested, the driver should provide a quiet ride.
- The fare should be clearly displayed on the cab's meter.
- The inside and outside of the cab should conform to road safety standards.

5 Car share

Most urban areas have car-share companies that allow you to use a car from their fleets for an hour, a few hours, or a few days. You join as a member for a nominal fee, then pay for use of the car on an hourly or daily basis. Gas and insurance is covered by the car-share company. By limiting yourself to a few hours or half a day, you learn to become very efficient in your use of the vehicle. People who give up their cars for car share usually save a lot of money not only on gas but also on insurance, parking, and maintenance costs.

> "[For] the six months that I don't plan to have a car, that adds up to $4,584 after-tax dollars in my pocket."
>
> Fred Langan, *Financial Post*, October 6, 2007, after doing the math on stopping his car lease and switching to a car-share program

Many of the car-share companies offer compact, fuel-efficient cars, as well as hybrids and diesels, so they are very good on gas and have low CO_2 emissions.

> "For every 1 new car that we put on the road for our members, there are 10 older cars that are pulled off of the road."
>
> Kevin McLaughlin, president of AutoShare, Toronto

FIVE BIG BENEFITS OF CAR SHARING

According to www.eartheasy.com/live_car_sharing.htm:

1. **Convenient.** There's almost always a car when you need one, and some programs have various models to suit your changing needs. Some motorists use car share as their second car.
2. **Practical.** Car sharing reduces the number of cars on the road. Each car-share vehicle replaces four to eight privately owned cars.
3. **Environmental.** Car sharing improves air quality by reducing individual car usage by as much as 50%.
4. **Low cost.** You pay only for the hours you drive.
5. **Carefree.** There's no maintenance, no servicing, no cleaning, and fewer parking hassles.

Check out these useful websites—they've helped many travellers across the country save money while they "go green" to get where they're going.

www.carsharing.ca: The go-to site for those new to car sharing.

www.zipcar.com: One of the first car sharing companies. Currently available in only Toronto and Vancouver, Zipcar is increasing its number of locations.

www.carsharing.net: Provides car sharing locations around the world as well as in Canada.

www.autoshare.com: Available in Toronto.

www.cooperativeauto.net: Available in Vancouver.

6 Rent a car

Instead of buying a massive fuel hog of a car for that occasional big-car need, consider renting one instead. You can buy a reasonably sized, fuel-efficient car for your daily needs and then rent the big one for the odd weekend or family holiday. The total monthly cost to you is much cheaper, and your gas bill will be slashed dramatically.

> "Average monthly lease payment for a midsize, 4 cyl car (not incl. gas and ins.) = $475
>
> Cost of renting a car for 1 solid week and 1 individual weekend per month (incl. gas and ins) = $375."
>
> Car$mart Inc.

You might also consider not owning a car at all and renting one when you need it. Much like car sharing, you'll be more frugal when you are paying for a rental car. These days, rental-car companies are offering fuel-efficient vehicles, and many have fleets of hybrids from which to choose. If you can manage it, rent during the week—the rental rates are lower than on weekends, and the price of gas when you refill the tank may also be lower (see Best Times to Fill Up Your Car, page 43).

7 Get a scooter

Instead of buying a second small car for your household, why not get a scooter? Scooter sales are skyrocketing these days as urban dwellers see the benefits of these two-wheeled machines. They are very fuel efficient, cheap to insure, and easy (and quite often free) to park. Gas-powered scooters require a special motorcycle licence to operate, but electric-powered scooters (or E-bikes) do not. This makes the electric version an easy and afford-able option—no licence or fuel required!

> "Electric scooters use 0 gas and can cost as little as $599 + tax, gas scooters use only 3.1 L/100 km on average and can cost as little as $1,149 + tax."
>
> Daymak Edmonton

Not only does a scooter save you money, it's cool and styl-ish. You can dress up this funky two-wheeler and travel first class. Here are five ways to jazz up a scooter:

1. Opt for a colour-matched helmet.
2. Match the colour of the travel box to that of the scooter.
3. Add on a windscreen: Not only does it reduce rain and bugs flying in your face as you ride, but it makes the scooter more aerodynamic.
4. Carry a contoured travel knapsack.
5. Buy a custom raincover.

MAKING SENSE OF TWO-WHEELERS

So, you want to turn in your four-wheeler for a two-wheeler? Or maybe you want one so that you don't have to buy a second car. Either way, two-wheelers use less fuel than four-wheelers, are easy to park, and can be a lot of fun to ride. There is a lot of choice out there; here's a breakdown of the motorized two-wheelers available:

- **Motorcycle.** This full-sized, motorized bike can cruise in both the city and on the highway. A motorcycle is powered by a gas engine, and engine size is usually 500cc or bigger. You'll need a motorcycle licence and insurance to operate one.
- **Scooter.** This motorized bike is also powered by gas only. Scooters are much smaller than motorcycles, and so are their engines. In fact, the engines are so small that these bikes are called "limited-speed motorcycles." Most are not allowed on highways. Scooters require a motorcycle licence and insurance to be driven on the roads.
- **Moped.** This too is a motorized bike but even smaller than the scooter. It has a gas engine, but it's very small. The moped also has pedals so that the rider can use muscle power to propel the bike. Less powerful than scooters, mopeds are allowed only on slow-speed roads. Because of the gas-powered engine, they require a motorcycle licence and insurance.
- **E-bike.** The e-bike (or electric scooter) is a new development in the motorized bike world. It's identical to a moped, but the engine is electric (and tiny) rather than gas-powered. It has pedals, so you can also use muscle power to move this machine. No motorcycle licence or insurance is required, and these bikes can be operated on the road or wherever bicycles are allowed.

8 Join a carpool

For many commuters, carpooling is nothing new. But are we all actively trying to carpool to save fuel? The typical car can comfortably seat four people, and a minivan can seat seven. Think of how much gas would be saved if we shared the journey with our neighbours. But carpooling is not just about commuting: We can save fuel by applying the carpooling concept to lots of other activities.

> "An average van pool of seven passengers emits about 7.5 times less pollution per kilometre than drive-alone commuting."
>
> The AutoSmart Guide, Natural Resources Canada, 2007

Here are five ways to incorporate carpooling into your daily routine:

1. Pair up with another family to go to your children's school recital.
2. Pile into one car for an office offsite meeting and share the expense.
3. Share the fuel and parking costs with your friends when you go to a sports or entertainment event.
4. Find a neighbour who does his or her grocery shopping on the same day as you and link up.
5. Pick up three of your child's teammates for soccer games and practices. Stay at the field to avoid driving back and forth. Watch them play, or read a book, listen to music, or snooze while you wait.

9 Telecommute

Like carpooling, telecommuting is not new. Yet, with today's advanced technology at our disposal, why are our roads so crowded with people driving to work? Our employers and governments need to offer incentives for those of us who want to work from home.

> "If everyone who could took full advantage of telecommuting, the reduction in miles driven would save $3.9 billion per year in fuel and the time savings would be equal to 470,000 jobs."
>
> 2005/2006 National Technology Readiness Survey, University of Maryland/Rockbridge Associates

Most people have computer access, and long-distance phone rates are at an all-time low, so there are usually no major costs involved in telecommuting. In this modern world, there's no need for employees to drive their cars for hours just to sit in an expensive cubicle at an office. See if you can work from home on a part-time basis or even one day a week. Think of the fuel you will save, how relaxed you will be, and how much pollution will be avoided.

10 Participate in no-drive days

Here's a chance to challenge yourself. Try to compress all of your car use into, say, Tuesdays, Thursdays, and the weekend, and leave your car parked on Mondays, Wednesdays, and Fridays. You could use public transit or a carpool, or work from home on these three days and commute in your car on the other two. Or go car-free on the weekends. If none of this fits your hectic schedule, start small and try leaving the car at home for (at least) one day per week.

> "The vision is that every Sunday in the month the world stops driving. Think about how much fuel we would save and the CO_2 reduction."
>
> www.globalnodriveday.com

The typical midsize car uses about 4.6 litres of gas and spews out on average 11 kilograms of CO_2 for each day that it's on the road, according to Natural Resources Canada's 2008 *Fuel Consumption Guide*. But this is based on a small, four-door car with a four-cylinder engine. Imagine the fuel used by a V6, minivan, or pickup truck. Keeping the car at home not only helps your pocketbook, it improves the air that you breathe.

11 Limit your use

All the tips so far have focused on alternatives to using our cars in our daily lives. But when we're in a rush or feel stressed, it's easy to give in to the car sitting there in the driveway, beckoning us to get things done "now." Stop and question yourself before you jump in the car and go. With fuel prices this high, it's time to curb the urge and limit the use of this polluting gas hog.

Consider these questions—and suggestions on how to keep the car parked and save fuel—the next time you're tempted to jump in the car:

Do you really need to pick up the dry cleaning right now? Why not wait until later when the roads are clear so that you don't get stuck idling in traffic? Can you combine this trip with other errands? Or does your dry cleaner have a delivery service?

Do you really have to dash off to the copy shop right this minute to copy those documents? Is it something that can be done at the office next time you're there, or at your partner's office? Or when you walk the kids to school, ask the school secretary if you can use the school's copier.

Is take-out coffee really that good or necessary that you have to drive for it, then sit in the drive-thru lineup idling your car? Why not brew coffee at home? You'll save on both the price of coffee and the price of gas, and you'll have more time to enjoy it, too.

PART 2

Fuel-saving maintenance for your car

12 Service, service, service

Properly servicing your car is key to its performance and longevity. But most importantly, you'll save a lot of fuel over your car's lifetime if you treat it well.

> "A poorly maintained vehicle can cost the equivalent of up to 15 cents per litre more on fuel each time you fill up."
>
> *Fuel Consumption Guide*, Natural Resources Canada, 2008

If you drive an average of 22,000 kilometres per year, you should have your car serviced at least twice a year, and even more than that if you drive a lot. In addition to an oil and filter change each servicing, the main items to monitor are tires, brakes, and air filter. Yes, this work can take time and cost a lot of money, but it's important for the fuel efficiency of your car and its resale value down the road.

Did you know that having your car properly serviced also promotes clean air? Ontario's mandatory Drive Clean and British Columbia's AirCare inspection programs, for example, help reduce airborne chemicals and smog-causing pollutants by identifying problems and having them repaired. All cars, trucks, and buses of a certain age must pass the inspection and be repaired if problems are identified. Plans are being developed to introduce similar programs in other provinces. So, keeping your vehicle in good shape will also help your lungs— and improve the air quality for everyone.

SERVICE GUIDE FOR FUEL EFFICIENCY

To maximize the gas mileage in your car, you need to look after it mechanically. Your car's owner's manual will outline the service work required. The chart below will help you get the best fuel efficiency out of your car.

Every Month	Every 6 Months	Every Year	Every 2 Years
• Check tire pressure	• Oil and filter change	• Replace air filter	• Check fluids: coolant, brake, transmission, power steering
	• Check air filter • Rotate tires	• Check brakes	• Check fuel injectors • Check emission control system (ECS)

Remember, always consult your dealership or local garage for the complete list of parts to be checked and maintained. Items such as turn signals, horn, and door mouldings don't affect fuel economy but are essential to a safe and properly functioning vehicle.

13 Choose synthetic oil

Synthetic (factory-made) oil is an alternative to petroleum (natural) oil. It's been around for a long time and can improve your vehicle's performance. Because it is engineered, it's contaminant-free and customized to work perfectly inside a car's engine. Synthetic oil molecules are light, smooth, and similar in size, so they slip and slide over one another with ease. If oil isn't moving easily in your engine, the heavy metal parts inside will bog down (increasing fuel consumption) and wear out (leading to costly repairs). The engine parts will move much more freely with synthetic oil than natural oil, particularly in extreme weather, either hot or cold.

> "[Advantages of using synthetic oil include] better cold driving characteristics, increased fuel mileage of nearly 10%, noticeably lower operating temperatures, better heat dissipation capability, and long-term high temperature stability."
>
> Don Stevens, *Making Sense of Synthetic Lubricants*

Synthetic oil costs a bit more and can be hard to find, but it lasts a lot longer than natural oil and so doesn't need to be changed as often. And the fuel savings and reduced engine wear are a big bonus.

14 Check fluid levels

Whether or not you switch to synthetic oil, you must still check your car's fluids periodically. Failure to do so may cause severe damage to the engine. But on a day-to-day basis, dirty or low fluids may cause the car to burn up more fuel than necessary. The fluids in your car are

- oil
- coolant
- brake
- transmission
- power steering
- window washer

The owner's manual will tell you how often these vital fluids need to be checked. Although you may need a mechanic for some, it's a good idea to know how to check the oil and the coolant yourself. These two fluids are the lifeblood of your engine, essential for its well-being. If the oil is low or looks brown or black, it's time to change it. If the coolant looks low or dirty, it needs to be changed, too. If these two fluids are fresh and clean, your engine can operate smoothly, without choking and sputtering and using extra gas.

15 Inspect the air filter

Your car's engine gets its power from mixing air and fuel and igniting the two in the cylinders with a spark plug. The air filter's role in this process is to make sure the air is cleaned before it meets the fuel. If the filter gets clogged with dust and dirt, the engine has to labour and strain to pull that air in for the right mixture. So a dirty filter makes your engine work harder, and it will burn up more fuel to do so.

> "Replacing a clogged air filter can improve your car's gas mileage by as much as 10%."
>
> www.fueleconomy.gov

To illustrate this process, imagine going for a jog or playing a sport with a dirty ski mask over your head. Surely, you would find it difficult to breathe! Usually, the air filter is easily accessible in a car's engine bay, so you can inspect it yourself (check the owner's manual before getting started). Pop out the filter. If it is dirty or dusty, it's time for a change. Typically, the air filter needs changing once a year. A dealer or garage can do this easily for you, or you can buy one and do it yourself.

16 Check the emission control system (ECS)

Have you ever had the "check engine" light on the dashboard come on? That yellow warning light has caused many a driver to panic. Quite often, the light appears because you have just filled up the gas and put the gas cap on incorrectly. This is an easy fix, but it may require going to a dealer or garage to get the light reset. However, if you didn't fill up and the light is on, it may mean that the emission control system (ECS) is acting up. The ECS senses how much air is being let into the engine and regulates the exhaust emissions coming out of it. If the ECS is not functioning properly and not allowing air in or out of the engine, it can put a serious strain on the engine.

> "Fixing a serious maintenance problem, such as a faulty oxygen sensor, can improve your mileage by as much as 40%."
>
> www.fueleconomy.gov

Always take the "check engine" light seriously. If it comes on, get your engine checked out immediately. Better yet, have the ECS tested each year to ensure that your engine is okay—before the light comes on.

17 Keep the fuel injectors clean

Like the air filter and emission control system, the fuel injectors are a key factor in your engine's performance. The injectors are basically a nozzle and a valve that shoot a perfect blend of gasoline and air into the car's cylinders. If they are dirty or not operating properly, fuel will get burned incompletely and the engine will malfunction.

> "Letting [the fuel injectors] become plugged leads to increased emissions and fuel use."
>
> *The Road to Green: Making the Shift to Eco-Driving*, Canadian Automobile Association

It's complicated to inspect or repair a car's fuel injector system. Leave this work to the professionals at your local garage or dealership. Or add a cleaning solution to the gas tank. This inexpensive liquid can clear out the gunk that gums up the injectors. No matter which approach you take, keep an eye on this vital part of your engine's operation—it will save you a lot of fuel in the long run.

HOW TO CALCULATE YOUR CAR'S FUEL EFFICIENCY

In Canada, fuel economy is calculated using L/100 km, or litres used per 100 kilometres:

1. Fill up from empty, noting the number of litres of gas you put into the tank.
2. Reset the trip odometer.
3. Record how many kilometres you travel on that one tank of gas.
4. Divide the number of kilometres by the number of litres you put in, then multiply the number by 100 to get the car's L/100 km.

For example,

60 litres into tank

÷

500 kilometres travelled × 100

= 12 L/100 km

The lower the L/100 km, the better.

For the most accurate measurement of your car's efficiency—to see how your car really performs—do this test at least five times, then average the L/100 km.

18 Inflate the tires

The tires are all that connect your car with the road. For your safety, it's crucial that you check them regularly (see Tip 20). Many of the cars on the road today are driving with under-inflated tires. Not only is this unsafe and damaging to the tires, but it can burn up a lot of extra gas.

> "For every 28 kilopascals (4 pounds per square inch) of under-inflation, fuel use increases by about 2%."
>
> *Fuel Consumption Guide*, Natural Resources Canada, 2008

This savings may not sound like much, but 28 kilopascals is a very small amount of air pressure. Imagine that you are down 28 kilopascals in each of the four tires—that's burning up 8% more fuel! Buy a tire-pressure gauge and check each tire, including the spare in the trunk, once a month. The right amount of pressure is stated in the car's owner's manual, or on the side of the tire or on a plate on the inside of the car's door frame. Fill the tires to the maximum listed pressure, but don't overfill, as this can be dangerous.

19 Add nitrogen to your tires

To minimize air loss in your car's tires, fill them with nitrogen instead of air. Nitrogen is a dry, inert gas that has larger molecules than air. This means it's more difficult for the molecules to diffuse through the side of the tire—so tires with nitrogen in them will stay pumped up longer than those filled with air. A properly inflated tire means better fuel economy and less wear (see Tip 18).

> "Nitrogen only costs $10 per tire and takes just 20 minutes to install."
>
> Peter Geer, service manager,
> 401-Dixie Nissan Infiniti, Mississauga, Ontario

Nitrogen is also cooler, dryer, and lighter than air, so your tires will roll easier, again using less fuel. You'll still need to check the tire pressure every month, but you will find that, with nitrogen, it will be more consistent. If the pressure of one of the tires is low, you can still add air to top it up—air and nitrogen can mix safely. Many garages and car dealerships are now carrying this great new product and can add it to your tires.

20 Buy better tires

It is important to regularly check the condition of your car's tires. Inspect all four tires to see if there are rocks or bits of glass or metal imbedded in the treads. If they look worn or cracked or make a whirring noise when you drive, it's probably time to replace them. Most garages or tire experts will give an assessment free of charge. Worn tires don't roll properly and resist the engine's energy; to fight this resistance, the engine will need to use more gas.

> "A 10% reduction in rolling resistance will result in a 2% reduction in fuel consumption."
>
> The Auto$mart Guide, Natural Resources Canada, 2007

Rolling resistance is the latest buzzword. The lower the resistance, the better the fuel economy. So when you shop for new tires, make sure they are rated for long life and have a low rolling resistance. All-season radials are usually the best. Remember to rotate the tires at least twice a year to maintain even wear.

21 Use snow tires if you need them

If you live in a snowy region, putting snow tires on your car in the winter will help immensely, especially if the roads are covered in snow most of the time. The snow tires will give you excellent traction, braking, and lateral stability. All this extra grip gives you better fuel economy because the engine doesn't have to work so hard to fight the slippage.

> "The cost of four high-quality snow tires mounted onto four steel rims ranges from $585 to $995. The cost of four high-quality replacement all-season tires mounted onto existing car rims ranges from $400 to $850."
>
> Jim Bridgeman, Steelcase Tire, Markham, Ontario

At some point, you'll have to replace your car's tires, so why not get the snow tires now? They are almost the same price as the all-seasons, but you get all the safety and fuel economy benefits right away. When you buy snow tires, make sure that you also get rims to go with them. This makes it easier (and cheaper) for the yearly swap of tires. Rims also help keep the tires round and prevent damage in storage.

WINTER TIRE SAFETY TIPS

Transport Canada, on its website www.tc.gc.ca/roadsafety/
safevehicles/safetyfeatures/wintertires/index.htm, provides the following winter tire safety tips:

- Install *four* winter tires.
- Buy tires marked with the symbol of a peaked
  mountain with a snowflake. This indicates that the
  tires meet specific snow-traction performance
  requirements and have been designed specifically
  for use in severe snow conditions.
- Don't mix tires with different tread patterns, inter-
  nal construction, or size.
- Don't use tires worn close to the tread-wear indica-
  tors.
- Check for proper air pressure to extend tread life,
  improve safety, and reduce fuel consumption.

If you live in an area where all-season tires just aren't
an option, or if you travel extensively throughout the
winter, snow tires are one of the most important
purchases you can make. For more information, check
out www.betiresmart.ca—a detailed website that gives
you the goods on different snow tire brands and proper
installation and maintenance. You can also stay on top of
emerging tire trends all year with their free monthly
e-newsletter, *The Tire Monitor*. Keep yourself and your
family safe this winter with proper tires.

22 Don't use snow tires if you don't need them

If you live in a moderately snowy region, you probably don't need snow tires. Snow tires are great when the roads are covered in snow, but they are not so great when the pavement is bare. They are made of a softer rubber than regular all-season tires, so they wear down easily on dry roads. The worst part is that snow tires are very knobby, with deep treads, so they are harder for the car's engine to turn. This makes the engine work harder and burn up more gas. In addition to the wear and reduced fuel economy, snow tires are noisy on dry pavement. They also don't grip as well on dry roads as all-seasons, so it takes longer for the car to stop. And who wants to lug these heavy, dirty things out of the basement or garage in the fall, only to have to put them back there in the spring?

> "Tires marked 'M + S'—or 'mud and snow' tires [are] also known as 'all-season' tires."
>
> Transport Canada, www.tc.gc.ca/roadsafety/safevehicles/safetyfeatures/wintertires/index.htm

23 Check the alignment

Your car's alignment is vital to your safety and its fuel economy. If the alignment is off, your car could crash. Have you ever experienced the feeling that your car wants to turn when you have it pointed straight? It can be quite alarming. But it also can be quite common. All you have to do is bump a curb or drive over a pothole to knock the car's steering out of alignment. This happens all the time, yet many of us aren't even aware that it has happened. But a misaligned car can burn up your tires and your gas.

> "Improper alignment will reduce your car's gas mileage because you will need to use extra energy to help keep the car in a straight line."
>
> www.dailyfueleconomytip.com

That extra energy comes from your engine—not your hands tightly gripping the steering wheel to maintain control! It's difficult to measure how much extra gas is used, as it depends on how severe the misalignment is, but if your tires are resisting the natural motion of rolling straight, you're burning additional gas.

PART 3

Great
gas tips

24 Don't install fuel-saving devices

Some companies try to convince consumers to buy their engine add-on devices or additives that will supposedly improve a car's fuel economy. These gizmos can be attached directly onto the engine or poured into the fuel tank. Be cautious. The Environmental Protection Agency and the Federal Trade Commission in the United States have each tested a number of these items and found that either they don't work or the fuel savings are grossly exaggerated.

> "Only six devices [out of 100 tested] indicated a very small improvement in fuel economy."
>
> www.epa.gov/otaq/consumer/reports.htm

What about the cost of these devices? And what happens if they damage your car? The auto manufacturer may not warrant the repair if you have tampered with the engine. If these fuel-saving claims were true and if it were so easy to do, why aren't these systems already incorporated into today's modern vehicles? A big "buyer beware" sticker belongs on any gadget or goo that claims to improve your car's mileage.

25 Install a fuel-use gauge

Despite my urging in Tip 24 to avoid all fuel-saving devices, this one is different, and it works. Installing a real-time gas consumption meter is a great idea. This is a computerized gauge that gives you an instant reading of how much fuel your car is using. It's easy to install yourself, or a mechanic can do it for you for a small fee.

> "Tank-to-tank monitoring of your fuel consumption isn't good enough. You need instrumentation that lets you reset the readout at will so you can track individual trips, or even portions of trips you regularly travel."
>
> www.ecomodder.com/forum

A fuel-use gauge keeps you honest in your driving. When you rev the engine or speed, it shows you how much fuel you are wasting. Keeping an eye on your digital "score" keeps your foot off the gas pedal.

26 Search for deals on gas

There are many ways to find bargains on gas, and they don't all require driving around in your car late at night. Shopping online for the cheapest gas station near you is an excellent way to plan your next fill-up. GasBuddy.com is a good place to start. It's an online service that tracks gasoline prices from coast to coast in Canada and the United States. Just select the area where you live and within seconds the website will tell you which stations have the highest and lowest prices.

> "Gasoline prices change frequently and may vary by as much as 20% within only a few blocks. It's important to be able [to] find the service station with the lowest priced fuel."
>
> www.GasBuddy.com

Get savvy and shop around for fuel (but not while driving in your car!). There's no need to be loyal to a brand just because it's at the corner of your street. Most gas comes from the same refining source, and the quality is similar at most chains, so go wherever the price is lowest.

BEST TIMES TO FILL UP YOUR CAR

Gas prices at the pump seem to fluctuate daily. There are several theories on when or where it is best to fill up on gas:

- **At night.** When the masses have all gone home and the roads are clear, prices at the pump tend to drop. There doesn't seem to be any real pattern to this, but it happens quite often. When the weather is hot, it's an especially good idea to buy gas late at night or early in the morning, since gas expands in heat, creating bubbles in the pumps. When gas is cooler, you will get more in your tank. On a hot day, you will get less gas and more bubbles.
- **Midweek.** People tend to use their cars a lot on the weekends, so the price is higher leading up to, during, and just after the weekend. In the middle of the week, the price of gas tends to drop, as there is less demand for it.
- **Before major holidays.** Three days before a major long weekend or holiday is a good time to fill up, as it's the calm point before a major peak. Gas stations tend to crank up their prices to capture motorists when auto travel is expected to be high.
- **Immediately after bad news.** When a war breaks out or there's a big hike in oil prices or a huge weather disaster, fill up immediately. Usually the day after such news breaks, the price of gas skyrockets. Not only does it go up, but it stays up for some time—even once things settle down.
- **Away from the highway.** If you have a long trip planned for the highway, try to avoid filling up at major exits. These main stops can hold you hostage with high gas prices. Fill up in town before you leave or exit at less popular spots to find cheaper gas.

27 Take advantage of loyalty discounts

Most gas companies offer some sort of loyalty incentive, usually through a points system. Fill up your tank with gas and you earn points that are good for use on items at the station's convenience store or for gifts online, including gas and gift cards.

> "You'll earn 1 or more points for every dollar that you spend."
>
> ESSO Extras program 2008 catalogue

These points systems may not be a huge payback, but they're better than nothing. Some large grocery chains and big box stores also offer discounts and loyalty points. Canadian Tire gives you cash bonus coupons (good old Canadian Tire money) when you pay for your purchase in cash, which you can then use at its retail stores and service stations. Loblaws and its affiliate grocery stores give you 2% cash back when you buy fuel at its Refuel gas bars. Many credit card companies are aware of consumers' desire to save on fuel and are now offering credit cards that link your purchases to the gas companies' points programs. These reward points can accumulate fast.

Your total yearly cost of fuel rises directly with the increase in the cost of fuel at the pump. It seems that these increases happen all the time. If you drive a midsize car, you'll probably use up 2,000 litres of fuel each year, according to Natural Resources Canada's 2008 *Fuel Consumption Guide*. This means that if the price of gas jumps from, say, $1.00 per litre to $1.40 (which can happen quite easily), you'll pay an extra $800 per year. If you have two cars—for instance, a midsize car and a minivan or SUV—this price hike will be well over $2,000 per year.

COST OF FUEL

				Cost ($)/Litre			
Litres	$0.80	$0.90	$1.00	$1.10	$1.20	$1.30	$1.40
800	640	720	800	880	960	1040	1120
900	720	810	900	990	1080	1170	1260
1000	800	900	1000	1100	1200	1300	1400
1100	880	990	1100	1210	1320	1430	1540
1200	960	1080	1200	1320	1440	1560	1680
1300	1040	1170	1300	1430	1560	1690	1820
1400	1120	1260	1400	1540	1680	1820	1960
1500	1200	1350	1500	1650	1800	1950	2100
1600	1280	1440	1600	1760	1920	2080	2240
1700	1360	1530	1700	1870	2040	2210	2380
1800	1440	1620	1800	1980	2160	2340	2520
1900	1520	1710	1900	2090	2280	2470	2660
2000	1600	1800	2000	2200	2400	2600	2800
2100	1680	1890	2100	2310	2520	2730	2940
2200	1760	1980	2200	2420	2640	2860	3080
2300	1840	2070	2300	2530	2760	2990	3220
2400	1920	2160	2400	2640	2880	3120	3360
2500	2000	2250	2500	2750	3000	3250	3500
2600	2080	2340	2600	2860	3120	3380	3640
2700	2160	2430	2700	2970	3240	3510	3780
2800	2240	2520	2800	3080	3360	3640	3920
2900	2320	2610	2900	3190	3480	3770	4060
3000	2400	2700	3000	3300	3600	3900	4200

Source: *Fuel Consumption Guide*, Natural Resources Canada, 2008

Here are some of the loyalty discounts offered by major gas retailers in Canada. The points may not seem like much at first, but they really add up.

Company	Program Name	Scale	Rewards
Chevron	Chevron/Texaco VISA	N/A	10-cents-per-gallon fuel credit rebate on gas purchases at Chevron and Texaco stations, 3% fuel credit rebate on any other purchases at Chevron/Texaco, and 1% fuel credit rebate on all other purchases
Esso	Esso Extra	$1 = 1 Esso Extra point	HBC discounts, RBC discounts, gift cards (Tim Hortons, Cineplex, HMV, Chapters, Roots), hotel discounts
Husky/Mohawk	Husky/Mohawk MasterCard	N/A	1% cash back on all purchases at any Husky, Mohawk, or Husky House restaurant location, and 0.5% cash back on all other purchases made with the Husky/Mohawk MasterCard
Olco	Participant in Gold Points Reward Network	N/A	Gold Points cardholders can collect points every time they visit a participating Olco station, with double the points when the card-holder fills up at Olco on Thursdays

Company	Program Name	Scale	Rewards
Petro-Canada	Petro-Points	$1 = 10 Petro-Points	Fuel savings, gift cards, and discounts on travel and vacations
Pioneer	Pioneer Bonus Bucks	$1 = 1 Pioneer Bonus Buck	Discounts on the purchase of fuel, car washes, automotive products, and everyday convenience items
Shell	Shell Escape	1 litre of fuel = 1 Shell Escape point	Ferrari merchandise, appliances, luggage, and fuel and auto service discounts
Sunoco	Sunoco MasterCard	N/A	4% discount on all Sunoco purchases, discount on NASCAR merchandise and events, and up to 1% discount on all other purchases
Ultramar	Ultramar MasterCard	N/A	Up to 2.5% cash rebate on all purchases at Ultramar, and up to 1.25% cash rebate on all other purchases

28 Use the right fuel for your car

Most cars run on regular unleaded gasoline, so there is no need to pay more and put mid-grade or premium gas into the tank. Many people buy premium, thinking that they are helping the engine. But they're not. At about 10 to 15 cents more per litre, buying up is just a waste of money. The upgraded fuels are of no benefit to your engine or fuel economy, so stick with regular.

> "Most **engines** run just fine on **regular** gas, which has an **octane** rating of about **87**."
>
> *2009 New Car Preview*, Consumer Reports

> "**High octane** fuels improve **neither** fuel **economy** nor **performance** and will just **waste** your money."
>
> www.greenercars.org

However, if your car requires premium gasoline (noted on the fuel-filler door or in the owner's manual), you must buy it in order for your car to run properly. When it runs rough, the engine will use up more gas and potentially have expensive problems. Premium fuel is required on most luxury vehicles and sports cars. The engines in these pricey machines require a higher level of octane to perform properly. An expensive car equals expensive gas!

29 Fill up near empty

Although waiting to fill up until the gas tank is near empty may sound like a risky proposition, filling up once it gets below a quarter-full can be a good way to maximize your car's fuel efficiency. Gas weighs a lot and slows down the car. This weight also taxes the engine, meaning it has to work harder, and burn more fuel, to keep you moving.

"[Keep in mind that every] extra 100 lbs (48 kg) of weight can increase your fuel bill by 2%."

www.eartheasy.com

If the price isn't right when your gas tank hits the quarter-full mark, put in just a bit of gas for the time being. You can fill up when you see a bargain on gas. The extra weight of a full tank will be offset by the cheaper gas. Some people call this "gas surfing." They believe that by no means should you let your car's gas tank get to the "light is on" stage—it's too risky. What if you can't find a station or if it's late at night and everything's closed? But, if you want to save money on fuel, gas surfing might be worth a try.

30 Don't overfill

Don't overfill your car's gas tank. When the pump clicks, stop. Don't round up to the next dollar. You're not helping your car by adding extra gas to the tank. If you keep pumping gas after the click, the excess fuel will spill down the side of your car or out through an overflow hole in the neck of the tank.

> "Topping off your gas tank is bad for the environment and your wallet."
>
> U.S. Environmental Protection Agency, www.epa.gov/donttopoff

Often when the pump clicks, the gas station's vapour-recovery system kicks into action, pulling back any of that extra gas you're trying to stuff into the tank. You've just paid for gas that ends up back in the station's main tank. Also, if you continually overfill, your car's own vapour-collection system may overwork trying to deal with the excess fuel and get fouled. Not only does this mean an expensive repair, but it will cause your vehicle to run roughly, which, in turn, burns up more fuel. It's clearly a costly mistake to top up.

31 Tighten the gas cap

After refuelling your car, make sure that the gas cap goes on smoothly and is tightened until it clicks. Two to three clicks are enough. If the cap goes on crooked or is loose, you'll lose fuel when it overflows out the neck of the tank as you drive. It could also evaporate out the gap.

"17% of the vehicles on US highways has either misused or missing gas caps, causing 147 million gallons of gas per year to vaporize into the atmosphere."

Service Tech magazine, September 2000

Tightening is one thing, but forgetting to put the cap back on is quite another. Sloshing fuel out the side of your car and not replacing the lost cap are both costly mistakes and potentially hazardous. In addition to losing fuel, your check engine light may also come on, another expensive thing to repair.

PART 4

Prepare
your car before
you drive

32 Lose some weight

Many of us could afford to lose a few pounds here or there, and so could our cars. Do you carry around a load of extra heavy items in your car? Get rid of the clutter on the back seat and in the trunk and lighten your load. Excess weight makes your car's engine work harder and burn up more fuel.

> "An **extra 45 kg** (about 100 lbs) can **increase** your **fuel** costs by **2%**."
>
> *The Road to Green: Making the Shift to Eco-Driving,* Canadian Automobile Association

This excess weight is particularly bad if you drive in a hilly area: Climbing steep hills with an extra load really uses up gas. Look to see if there's too much junk in your trunk. Extra tires, bags of sand for winter driving, heavy bathroom tiles, golf clubs, a big bag of dog food—all are unnecessary to drag around. Make it easier on your car and your pocketbook.

33 Reduce drag

Roof racks, cargo boxes, and trailer hitch attachments slow you down. The aerodynamic drag caused by these add-ons overworks your car's engine. Your car has to push much harder to force you through air, increasing fuel consumption.

> "A loaded roof rack can decrease your fuel economy by 5%."
>
> www.fueleconomy.gov

The weight of putting things in or on these attachments is another culprit. The drag from air is about 3% and the weight can be an extra 2%. Whenever you can, avoid putting objects on top of the car, storing them inside the car or trunk instead. If you drive a pickup truck, keep the tailgate down when driving on highway (assuming there isn't anything inside the box that could fly out). A pickup truck's box has incredible drag when the tailgate is up. By opening the gate, the air can flow freely over the truck, increasing fuel economy. Also remove any brush bars from the front of the truck. These metal bars have drag, can whistle, and add extra weight to the front end.

34 Keep the car's body clean and smooth

Make sure that your car isn't dented or missing pieces. Not only does it look bad, but your car won't cut through the air on the road as efficiently. Aerodynamic drag is a total drag on your fuel bill! (See Tip 33.) Keep your car washed and waxed, too. A smooth, clean car can use up to 5% less fuel than a dirty one. If you live in the country or near a lot of construction, check the underbody and wheel wells for caked-on dirt and mud. If you live in a winter climate, check these areas for built-up snow and ice.

> "During cold weather watch for icicles frozen to [the] car frame. Up to 100 lbs. can be quickly accumulated!"
>
> www.GasPriceWatch.com

This excess weight is a real drain on fuel. Also check the wheels. If they are bent or broken, they won't roll properly and will use up more gas.

There's another reason to keep your car's body free of damage: You could get pulled over by the police. In many regions it's an offence not to report damage to your vehicle, even something as minor as a dented bumper. If you have the misfortune to be in a minor accident and there is visible damage to your car—and you and others involved aren't injured—head to the nearest police station or collision reporting centre as soon as you can and report the incident.

35 Power down

Your car's electrical accessories and power items are a big drain on energy. This energy comes from the battery and the engine (the big gas burner). Limit the use of anything that requires power in your car. When you arrive back home, be sure to shut off all the power items—DVD screen, cell-phone chargers, portable GPS unit, iPod plug-ins, what have you. Left on, these items not only can drain your car's battery overnight but they will give the engine a huge shock in the morning when you restart the car.

> "Items that plug into your vehicle's cigarette lighter ... can cause the alternator to work harder to provide electrical current. This adds a load to the engine and added load increases fuel use, decreasing your gas mileage."
>
> www.eartheasy.com

In addition, before you start the car in the morning, don't plug in these accessories until the car has had a chance to warm up. Get rolling and warm up the car, then plug in (while stopped, of course) the equipment. It will save fuel, your battery, your alternator, and the strain on your car's electrical system.

PART 5

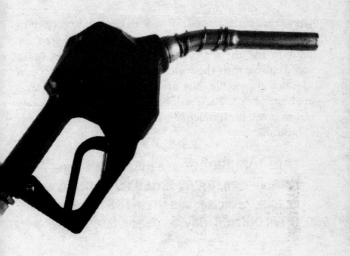

Avoid the
extremes of
hot and cold

36 Windows up or down?

On warm days, start your drive with the windows open to get rid of the hot air inside the car. This will lessen the need to use the air conditioner, saving fuel. At lower speeds, drive with the windows down and sunroof open as much as possible—city driving and stop-and-go traffic are best for this. There will be a bit of aerodynamic drag, but that will burn much less fuel than turning on the air conditioner.

> "Air conditioner use increases fuel consumption, increases NO$_x$ emissions in some vehicles, and involves environmentally damaging fluids."
>
> www.greenercars.org/drivingtips.htm

However, when you drive on the highway or are travelling at high speeds, keep the windows and sunroof closed. Open, they create a huge air pocket inside the car, forcing the engine to work much harder. At the same time, try to avoid using the air conditioning when-ever possible (see Tip 37).

37 Avoid air conditioning

Your car's air conditioning system is a real gas hog. It's actually one of the worst fuel-economy offenders. As much as possible, avoid using the air conditioning. Opening the vehicle's vents when driving on the highway or opening the windows and sunroof when driving in the city will help cool things down.

> "Air conditioning can increase fuel consumption by up to 20% due to the extra load on the engine."
>
> *Fuel Consumption Guide*, Natural Resources Canada, 2008

A car's air conditioning system requires an immense amount of energy to operate, straining your car's engine. Just listen to the noise the engine makes when you push the a/c button. Those are the painful sounds of excess fuel being burned up. If you have an auto climate system in your car, the air conditioning unit will be on all the time. Auto climate uses the air conditioner to modulate the air in the car. Override this by turning the a/c button to the off position. Another tip is to turn off the air conditioner about five minutes before reaching your destination. The car will continue to circulate the cool air inside for a short while once the system is turned off.

38 Chill in the shade

Whenever you can, park your car in the shade—under a tree, in an underground parking lot, or in the shade of a big building. A car can get up to 15 degrees Celsius hotter sitting out in the sun, leading to a number of problems.

> "Besides helping to keep your car cool, which reduces the need for air conditioning, parking in the shade also minimizes the loss of gas due to evaporation."
>
> www.eartheasy.com

This extra heat is difficult to get rid of, so you'll be more inclined to use the air conditioning to cool off. The heat can also cause the gas in your tank to evaporate right out the cap. If you can't find a shady spot, park the nose of the car pointing into the path of the sun to protect the gas tank from the direct sun. You may also want to buy foldable window shades to block out the sun's rays. Or you could tint the windows with a dark, protective film to permanently block out the heat of the sun. If you know that it's not going to rain, leave the windows open a crack to allow the hot air to escape from the car's cabin.

39 Use the garage

If you have a garage or have access to one, use it in extreme weather. In the winter, you won't have to clear the snow and ice off your car. The car will also get warmer much quicker after you start the engine if it's been kept inside. This puts a lot less strain on the engine and so saves fuel. In the summer, your car will stay cooler being in the shade of the garage, so you won't need to use the air conditioning.

> "Parking in your garage will help your car stay warm in the winter and cool in the summer, and you won't have to depend as much on your gas-guzzling air conditioning or defroster when you drive."
>
> www.bankrate.com/brm/news/auto/fuel-efficient/5.asp

An added bonus to keeping your car in a garage is that it won't rust or fade as quickly as it would if left outside, since it won't be exposed to the elements. So get rid of all that extra stuff you don't need in the garage and make room for your car.

40 Install a block heater

Using an engine block heater is an excellent way to get your car warmed up in the winter. It can be easily installed onto any vehicle, for a nominal charge. This device warms the engine block, oil, and coolant before you start the engine. For those cold mornings, plug it into an electrical outlet, set the automatic timer to provide heat two hours before you plan to leave in the morning, and that's it. When you start your car (and it will always start!), it will operate at its peak performance right away. No need for idling or revving the engine to get it going—you'll have heat in a matter of seconds.

> "Block heaters can **improve** overall winter fuel **economy** by as much as 10%."
>
> Fuel Consumption Guide, Natural Resources Canada, 2008

A cold metal engine uses more energy to get going than a pre-warmed one, so use a block heater and save on fuel.

Make sure that you're familiar with the various types of block heaters:

- **Dipstick heater:** Sits in your engine oil, replacing your regular dipstick—works by warming your oil.
- **Inline heater (circulating):** Spliced into the heater core hose—works by pumping coolant over its heating element.
- **Inline heater (non-circulating):** Spliced into the coolant hose.
- **External element:** Bolts on flat against the engine block.
- **External element (magnetic):** Sticks flat against the engine block or oil pan.
- **Frost plug heater:** Replaces your car's existing frost plug—works by warming the coolant inside the engine block.

41 Don't use a remote starter

Some cars have a remote start system that allows you to automatically start your car before you get into it. You just push a button on a key fob (even from inside your house) and your car will be warmed or cooled for you when you get into it. Although this sounds luxurious, it's all horribly wrong. This is a huge waste of fuel and not environmentally sound. A freezing cold car that is idling with the heat on or a boiling hot car that is idling with the air conditioning on uses a lot of fuel.

> ## "Idling is 0 miles per gallon."
> California Energy Commission, www.consumerenergycenter.org

It takes only a few minutes to cool a car by driving with the windows open and even less to heat a car with regular driving, so it's a huge waste of gas for such a small luxury. Remote starters can be expensive and can void your car's warranty if installed improperly. Avoid these wasteful gizmos.

42 Don't rev the engine

Whether it's cold or warm outside, never rev the engine when you start it. Your engine does not need any help in starting or warming up. Driving is the best way to get the car going—that is how it's designed. If you rev the engine, you are overriding the programmed start procedures of your car, which will end up burning a lot of fuel and could potentially damage the engine.

> "Revving your engine not only wastes gas but some gas may wash down your engine's cylinder wall and get into the oil pan, causing excess engine wear as well as diluting your engine oil."
>
> "Auto Facts & Tips," www.speedy.com

Particularly in winter, revving the engine is very bad. A cold engine is at its worst in terms of fuel consumption and emissions. As tempting as it may be, revving the engine causes havoc, so don't do it.

43 Wear appropriate clothing

In extreme weather, plan ahead and wear the right clothing for your car. In freezing cold weather, you can layer clothing, peeling it off as the car warms up. This will eliminate the need to idle your car to warm it, wasting gas. You can also wear a hat and gloves to help keep toasty. In hotter weather, wear lighter clothing to offset the higher temperature in your car. This will lessen the need to use the air conditioner to cool off. You may want to use a beaded seat cover, which puts a buffer between you and the car seat for increased air flow, helping to keep you cool. Although the beads may seem uncomfortable at first, they can be actually quite soothing once you get used to them.

> "**Prepare** yourself for **winter** driving. Choose **warm** and **comfortable** clothing. If you need to **remove** outdoor clothing later while driving, **stop** the vehicle in a safe spot."
>
> Canadian Centre for Occupational Health and Safety, www.ccohs.ca

44 Use heated seats

Many cars now come with heated seats. If your car doesn't have them, consider getting them installed or buying a heated seat cover. In addition to their obvious winter use, they are great for cooler mornings in the spring and fall. Despite the rumours, they don't weigh very much or use up a lot of energy, but they sure warm up your body quickly in the winter.

Average cost of factory heated seats = $250

Average cost of aftermarket installed heated seats = $450

Average cost of slipcover with heated coil = $150

CarSmart Inc.

This instant heat reduces the need to warm up your car by idling or revving it to get heat out of the vents, thereby reducing fuel consumption. There is an upfront cost to these heaters, but it doesn't take too many tanks of gas at today's prices to pay off.

PART 6

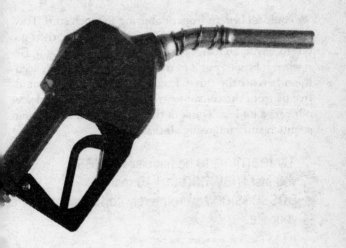

Plan your
drive

45 Be an anti-consumer

We could all benefit from not buying so much stuff. This stuff tends to rule our lives: We plan to buy it, then go out to buy it, exchange it because it's not quite right, fix it when it breaks, put it in the garage when we don't need it, and eventually throw it out. Having less stuff would free us from the consumer treadmill. And we would use our cars a lot less. Think of the savings from not buying so much stuff and using all that gas!

> "By learning to be happier with less, you just may find that so many possessions were merely complicating your life."
>
> *The Anti-Consumer Advocate*, Bob Horowitz, Sustainable Enterprises Inc.

Studies show that, next to commuting, shopping is the second biggest use for our cars. It's time that we all spent less time on the road. Think of what you could do with all that extra time. Consume less stuff, consume less gas—a huge savings all around. And a better, simpler world for all of us.

46 Use mail order or couriers

Mail trucks and courier vans are constantly driving around making deliveries to homes and businesses. Many of them run on diesel fuel, propane, or (coming soon!) electricity, so they are a more efficient way to move products than gas-fuelled vehicles. Why waste gas driving all over the place to do your shopping? A courier could bring it right to you. Many stores offer free delivery, so why not go online, purchase your goods, and have them delivered to your home?

> "Books, music and video—the shopping segment that began the online retailing movement in the 1990s—grew by a combined 32% to $4.14 billion in 2007 from $3.13 billion in 2006."
>
> www.internetretailer.com

Try to incorporate "delivery" into your life. Always be asking yourself if that item you need could be brought to you. Save time, save hassle, but mostly, save fuel.

THINGS TO HAVE DELIVERED OR DO ONLINE

There are hundreds of things that we do in our everyday lives that can now be done online. Many of the items we buy each week can be delivered, too:

- **Dry cleaning.** Why drive your dirty laundry over to the cleaners? You can easily and reasonably set up an account online, hang your clothes at the front door, and have them picked up and returned to your home.
- **Office supplies.** Need folders for your home office and a new chair for your kid's desk? Order them online from one of the many big box office-supply companies and they will be at your door within a day or two. Often the shipping is free.
- **Banking.** Who wants to drive all the way to the bank, find a place to park, and then line up to pay bills? Most banking can all be done online 24/7. There really is no need to ever set foot in a bank.
- **Grocery shopping.** Almost every major grocery chain offers an online delivery service these days. For many, shopping for food is a real chore, and it often requires a long journey in the car to get to the store. Order your weekly groceries online and have them delivered to your door—usually for no extra charge.
- **Gift delivery.** Buying a gift for a loved one can be exciting, and you may feel you need to drive to several shops before finding just the right one. Why not browse the shops online, then have the gift delivered to your home or theirs? A special delivery can have a big impact.

47 Plan ahead

Taking a few minutes to plan your trip will save fuel
and time. Here are 10 helpful ideas:

1. Use the internet or a GPS unit—or even a good old map—to plan
    your journey (don't waste fuel getting lost or taking the longest
    route).
2. Take a route that doesn't have hills, construction, or notoriously
    heavy traffic.
3. Take a route that doesn't have a lot of left turns (don't get stuck
    idling in a long line waiting for the light to turn).
5. Avoid rough roads that have dirt, gravel, or snow (you use up
    more energy on these and may damage your car).
5. Avoid running errands at rush hour.
6. Call a store or check online to make sure it's open before head-
    ing out in your car.
7. Call the store ahead of time to see if it has the item you want to
    buy.
8. Shop around ahead of time (don't immediately head out to see
    the deals; instead, price-compare online or in the newspaper
    first).
9. Do you really need to go right this instant? Rushing out usually
    means a wasteful journey—be patient and plan your errands so
    that you can do several at once. That item you want might even
    go on sale if you wait a week.
10. Do you really need to go at all? Maybe you need to rethink things.
    Remember: anti-consumerism!

48 Give yourself time

One of the biggest obstacles to driving efficiently is being late. When we rush, we often end up driving like maniacs. Racing from stop sign to stop sign puts a huge strain on the engine. The waste of fuel can be severe.

> "Aggressive driving (speeding, rapid acceleration, and braking) wastes gas. It can lower your gas mileage by 33% at highway speeds and by 5% around town."
>
> www.fueleconomy.gov

Plan to leave a little earlier so that there's no pressure on you to drive faster and race against the clock. This will also give you a buffer if you get delayed by an accident or construction. Driving your car with time on your side not only saves you money but also makes the journey so much more relaxing.

There are many other things you can do to improve the quality of your trip while you're factoring in the extra time. Think of it as "free time"—turn off your cell phone or BlackBerry and refresh your brain. If you have a long commute or you frequently get stuck in rush-hour traffic, consider listening to books on tape while you wait. At two hours or more a day, you can get through a library's worth in a year. Or, use the time to teach yourself another language by listening to taped language lessons. Extra time on the road really can be quality time—just make sure to keep your eyes on the road.

49 Avoid short trips

Short trips in your car waste a lot of fuel. Your car's engine needs to be warm to run efficiently. This means short, quick-stopping trips are not efficient.

"Trips of less than five kilometres generally do not allow the engine to reach its peak operating temperature (especially in cold weather), and that means fuel consumption and exhaust emissions will be significantly higher than when covering the same distance with a warm engine."

The AutoSmart Guide, Natural Resources Canada, 2007

When your engine is cold, it requires a lot of energy (which means burning gas) to move the metal parts and pump oil through it. When you take your car from a cold start (even in the summer) and drive it a short distance, it never gets a chance to warm up. If from there you head off on another quick errand, it again doesn't get the chance to warm up. Leave your car at home and use non-car means (see Tips 1 to 11) to get these short trips accomplished.

50 Try trip-chaining

When you need to run errands, combine several in a single trip—this is known as "trip-chaining." It's a much more efficient use of your time, with less gas wasted. Make sure to plan ahead (see Tip 47) and that your car is warmed up in an efficient way (see Tip 56) before heading out. By linking together seven or so trips instead of doing each individually, you'll be much more effective and efficient.

> "First, I get a lot accomplished in a short amount of time. Second, I don't take time to linger and lust after items I see in the stores. I'm on a tight schedule, and I don't have the time to browse. I stick to my list and get in and out as fast as I can. Finally, it frees up other days in my week ..."
>
> Lynnae, a "frugal mom" blogger, www.beingfrugal.net, February 12, 2008

In addition to the time and fuel savings you get from trip-chaining, you will put a lot less strain and mileage on your car, saving you money on repairs down the road.

51 Park 'n' walk

Instead of trip-chaining in your car, try driving to a central location, parking your car, and then walking, doing your errands on foot. It requires a bit of planning and you may have to pay for parking, but you'll save a lot of gas, you'll pollute less, and you'll get some exercise.

> "It's a good idea to ask yourself every single day, do I really need to take the car to get there?"
>
> Gillian Deacon, *Green for Life*

Or go to a nearby shopping mall: Since malls have a number of stores under one roof, there's no need to drive from shop to shop. When you get to the mall, avoid driving near the storefronts or main entrances—they are always jammed with idling cars. And don't drive around and around looking for that perfect parking spot. Park in the first empty space you come across and get walking.

There are added benefits to the park 'n' walk plan. You'll not only give your car a rest, but you'll feel better. If you add just 10 minutes of exercise to your daily routine, you'll get nearly an extra hour of exercise a week. And you'll notice the benefits, too. Better cardiovascular health will help you sleep better, feel more energetic, and live longer. So make the effort by parking that extra block or two away. It all adds up to improved health, and it will also be good for your blood pressure—you won't get angry at other drivers who steal the parking space you're waiting for.

52 Park 'n' ride

The catchy term "Park 'n' Ride" has been used by many major North American transit organizations. It refers to a system that allows suburban commuters to drive their cars to a lot near a public transit station or stop, park the vehicle on a lot for the day, and ride transit downtown. There are several benefits to this:

- Fuel and emissions savings from idling in congestion
- Stress relief from not having to navigate busy roads
- Parking savings from avoiding expensive downtown lots

Although there is a charge for the transit, it can be crowded, and the parking isn't always free, the benefits of Park 'n' Rides far outweigh these minor costs. Or, in a twist on the Park 'n' Ride concept, put your bike in your car's trunk (keep in mind Tip 33—don't put it on the roof!) and drive to an area close to the city where you can park for free. Then hop on your bike and pedal away—a free method of transportation with zero emissions and zero gas used.

53 Park with the hood pointing out

Parking with the hood facing out of a parking spot saves fuel. Not only will you be able to drive out with ease the next time you use the car, but all of the jockeying around to park with your hood pointing out is done when the engine is warm, so it's easy on both the engine and fuel—backing out and manoeuvring such a big, heavy machine when the engine is cold takes a lot of effort (and gas). Or, rather than backing into a parking spot, drive through the spot to the empty one adjoining it, so that you can exit the spot in a forward motion.

> "Fuel consumption and pollution output are much higher in the first minute or two after a cold start ..."
>
> The Auto$mart Guide, Natural Resources Canada, 2007

Avoid parking uphill. It will require a lot of energy to get the car going up that hill when you restart it later, when the engine is cold.

Both are little steps, but helpful in the grand plan to save fuel and reduce exhaust emissions.

54 Use the most efficient vehicle

The average American household has 2.28 cars. More than 35% of American households have three cars, according to a February 12, 2008, Experian Automotive study. It's no wonder we're running out of fuel. Quite often families with two cars have a small car that seldom gets used and a bigger, family car. For fuel savings, use the smaller car more frequently and leave the big car in the driveway.

> "Extra weight can really penalize you in stop and go driving, where the extra weight costs you money every time you start off from a stoplight."
>
> 2009 New Car Preview, Consumer Reports

Use the smaller car in the city, with its stop-and-go traffic, and save the family hauler for longer, highway drives. On the open road, the bigger car will have similar fuel economy to the smaller one, but in town, it will be significantly worse.

55 Give up your Sunday drive

Sorry to report, but Dad's Sunday drive in the family auto is just not a good idea anymore. Piling into the car and going for a cruise is a huge waste of fuel. Most times these drives involve a lot of aimless, stop-and-start manoeuvres that are really hard on the car. It's too bad that "it ain't like it used to be," but back in the old days, gas was cheap, the roads were clear, and a joyride was actually joyful. Now it's just plain expensive.

> **"25%** of Canada's **greenhouse** gas emissions come from transportation, and **half** of that amount comes from **personal** passenger use."
>
> Statistics Canada

It's high time that we think hard about our use of the car. Yes, the car is something to be proud of and can give us a sense of freedom and release, but every time we use it, it costs us. Our pocketbook, our air, and our water are all affected. Try to view the car as an appliance that you use only when necessary.

PART 7

Put an end
to idling

56 Don't idle

Idling is wrong for many reasons. It is excessive, wasteful, and horrible for the environment. A car's engine is designed and built to be running in a moving car, not while the car stands still.

> "Modern vehicle engines only require 30 seconds of idling to circulate the oil through the engine. Allowing your vehicle to idle for more than 30 seconds wastes fuel and does nothing to heat the engine."
>
> The Road to Green: Making the Shift to Eco-Driving, Canadian Automobile Association

When you idle the engine to get your car warm, all you're doing is burning up fuel and polluting the atmosphere with poisonous CO_2. As the saying goes, idling your car equals zero miles to the gallon.

> "Avoiding excessive idling can save up to 19% [of your tank of gas]."
>
> www.edmunds.com/advice/fueleconomy/articles/106842/article.html

Excessive idling can also damage the engine. The oil inside can break down and the metal parts start to wear if the engine is not working at its optimal performance level. In more ways than one, idling is incredibly harmful. We need to put an end to this bad habit.

57 Turn the engine off

Other than leaving your car at home, what is the best way to avoid using fuel? Turning off your engine. Idling is one of the biggest drains on fuel economy (see Tip 56).

> "If you are stopped for more than 10 seconds, except while in traffic, turn off your engine ... idling for more than 10 seconds uses more fuel than it takes to restart your engine."
>
> Fuel Consumption Guide, Natural Resources Canada, 2008

There are many myths about turning a car's engine off and on. Some people think that it takes 10 times as much fuel to restart a car than letting it idle. Others think that you cause great strain and harm on the starter and electrical system when you fire up the engine soon after shutting it off. Both are false. The goal is to save fuel and cut CO_2 emissions, so turn it off. If your car isn't idling, you aren't wasting fuel.

If you have to wait somewhere for 5 or 10 minutes, be patient and turn the car off. If you find yourself with more than 15 minutes and the temperature is extreme, seek out a place that's heated or air-conditioned. Think how much more pleasant it would be to wait in a bookstore or coffee shop—and be saving money while you're at it.

PLACES TO TURN YOUR ENGINE OFF

Idling is really bad: bad for fuel consumption and bad for the environment. The good news is that most people are aware of this and are making the effort to stop and turn off their cars rather than keeping them running unnecessarily. Here is a brief list of a few key places where you should turn off the engine rather than idle:

- **Schools.** In many provinces, it's illegal to idle near a school when dropping off or picking up the kids. Most people know this, yet many continue to do so, coming up with all sorts of excuses. None are acceptable.

- **Construction zones.** Getting stuck at a construction zone often means we sit in ours cars, idling, for many minutes. As soon as you are stopped (and see that there won't be any further movement for a while), turn the car's engine off. If you do, others will follow.

- **Long lights.** If you arrive at a major intersection and the stoplight has just turned red, you may be in for a wait. Instead of sitting there burning up fuel and polluting the air, turn off your car's engine; restart it just before it turns green again (when the light controlling the opposite traffic turns amber).

- **Stores.** If you are with a friend who needs to make a "quick" stop at an ATM or a convenience store, turn off the engine while you wait—often that quick stop ends up taking five minutes.

- **Accident sites.** Let's hope that you never get stuck at the scene of an accident. If you do get caught waiting in the jam of cars, turn your engine off, put on some gentle music, and relax. It may be a while before things get sorted.

58 Don't do drive-thru

To save fuel we should all avoid the drive-thru. Never mind that the fast food is likely not good for us anyway, even worse is the fuel that we waste as we idle our cars waiting in line.

"If every Canadian motorist avoided idling their vehicle for just five minutes per day, more than 1.4 million tonnes of CO_2 would be spared from entering the atmosphere and contributing to climate change."

The Auto$mart Guide, Natural Resources Canada, 2007

Studies show that the typical drive-thru wait is just over five minutes. That doesn't sound like much, but while you wait, your car wastes fuel and emits a huge load of CO_2 into the air. And drive-thrus aren't just about food anymore. These days, you can do your banking, pick up beer, fill a subscription, and drop off your dry cleaning from your car. Although drive-thrus are convenient and seem to save a lot of time, make the effort to park your car and walk into the store. Save your fuel, save your air, and save yourself from drive-thru idling.

PART 8

Gas-saving
driving habits

59 Stop speeding

Speeding is bad for numerous reasons. It's unsafe, it's against the law, and it's a huge waste of fuel. Cars are engineered to drive at standard speed limits. The limit varies by province, but generally, 92 km/h is the sweet spot for fuel efficiency in a car. Anything above this speed (no matter what the size of vehicle or engine) leads to a steep decline in fuel economy.

> "Decreasing your speed from 120 km/h to 100 km/h can reduce your fuel consumption by up to 20%.
>
> Fuel Consumption Guide, Natural Resources Canada, 2008

If you drive at 130 km/h, you will use 30% more fuel. The more you speed, the higher the percentage of gas. The reason for this is wind resistance. Forcing your car through the air at high speeds creates a huge strain on the engine. This leads to exponential fuel consumption. Try going the speed limit (it's the law, you know!) and watch your fuel gauge move much more slowly.

60 Use cruise control

One way to prevent speeding and save fuel is to use cruise control as much as possible when highway driving. This device sets your speed to an efficient, fuel-saving level and helps you avoid wasteful on-and-off acceleration.

"Up to 14% savings, average savings of 7% [by using cruise control]."

www.edmunds.com/advice/fueleconomy/articles/106842/article.html

Many cars these days have cruise control as a standard feature, but it can also be added to any car for a minimal cost. Cruise control is best used on flatter, open highways; in hilly or mountainous areas, the cruise control will continually tell the engine to rev up for the hills. In addition to saving your gas, cruise will save your driver's licence. Fixing your car's speed at the posted speed limit guarantees that you won't get into any trouble with the police. But remember to drive in the far right lane when using cruise control, as the rest of the traffic tends to move at a pace well above the limit. You may have to endure some honking horns and scowling faces when you drive at the speed limit.

61 Don't drive aggressively

Being in a rush can make us want to speed in an effort to make up lost time. But aggressive driving wastes fuel and will actually reduce travel time by only 4%, according to Eartheasy, an organization devoted to sustainable living. We've all seen someone racing from a traffic light, only to get stuck at the next stop. This "flooring it" approach never pays off.

> "Jackrabbit starts and hard braking can increase fuel consumption up to 40%."
>
> The Road to Green: Making the Shift to Eco-Driving, Canadian Automobile Association

The key to saving fuel is to drive smoothly away from a light or stop sign, coming smoothly to a stop at the next one. Thrashing away at the pedal makes your engine work hard to get the car back up to speed. Stomping on the brakes is equally bad, as it takes all the engine's effort and squashes it. Now your engine has to rev up and start all over. In addition to wasting fuel, this aggressive style of driving is hard on your car. Engine, brake, and tire wear all increase exponentially when you drive in this manner.

"LOG" YOUR FUEL ECONOMY

If you want to get serious about saving fuel, start a log. Keep track of your journeys, especially the ones you routinely travel. Aim to beat your record on these trips. Keeping a log makes saving gas a fun exercise, but it also may alert you to problems. If your results start to vary widely, you may have a problem with a tire, an air filter, or the fuel injectors.

Here's a log for tracking your fuel economy. A sample entry is included to give you an idea of how it could look—and how much you'll save.

Start	End	Distance Travelled	Fuel Used	Fuel Economy
(km)	(km)	(km)	(L)	(L/100 km)
S 20,938	E 21,643	$D = E - S$ 715	F 55	$F \div D \times 100$ 7.69

62 Drive steadily

Driving smoothly and steadily is a vital part of conserving fuel (see Tip 61). Squeeze the gas pedal off the line and squeeze the brakes to come to a smooth stop.

> "Flooring the gas pedal not only wastes gas, it leads to drastically higher pollution rates. One second of high-powered driving can produce nearly the same volume of carbon monoxide emissions as a half hour of normal driving."
>
> www.greenercars.org

Slowing down and speeding up take extra fuel. But since often we need to do so to keep with the flow of traffic, let's at least do so smoothly. Imagine a full bucket of water sitting next to you in the passenger's seat. One wild takeoff or panicked brake will knock that bucket over and spill water onto the seat and floor. Constant speed and momentum is key here: You want to keep the car rolling but not erratically. When you drive smoothly and steadily, the engine gets into a relaxed zone and burns fuel more efficiently. You'll get into that zone, too, and will be much calmer as you drive.

63 Lighten up on the brakes

This might sound odd, but try to limit your brake use. Your car's brakes are a big enemy to inertia. Every time you brake, you slow down the vehicle, which means the engine will need to exert a lot more effort to get it back up to cruising speed. The revving-up cycle after braking is a big waste of fuel.

> "Avoid coming to a complete stop whenever possible (and when it's safe and legal, of course). It takes much less energy to accelerate a vehicle when it's already traveling ..."
>
> www.ecomodder.com

Cruise or coast wherever you can. Look ahead and plan for stoppages. If you see a green light that is still far off, it will likely be red by the time you reach it, so just glide the car. Is there a stoplight at the bottom of a hill? Stay off the gas, roll down the hill, and coast up to the light. Not using the gas pedal is key to not using the brakes: Stay off the gas, and the brakes won't be needed.

PART 9

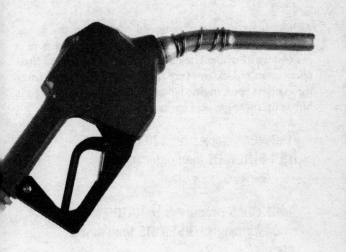

Choose a more fuel-efficient car

64 Buy a new car

Although getting rid of your old car and buying a new car is a big expense, the savings in fuel and repairs may make it worthwhile. The outdated technologies and tired parts in an old car are very inefficient. This inefficiency means that your engine labours, which in turn means it burns up more gas and emits more pollutants.

> "Newer vehicles tend to be more fuel-efficient than older models."
>
> www.oee.nrcan.gc.ca/transportation

> "Old cars produce 19 times more smog-forming emissions than newer models."
>
> "Time to Get Rid of That Old Beater?" Michelle Lalonde, Montreal Gazette, September 9, 2008

If it's time to make the move to a newer, more fuel-efficient vehicle, surf the internet before you venture out to a dealer showroom. Websites such as www.edmunds.com and www.consumerreports.org offer unbiased professional opinions on which cars are reliable, safe, and fuel efficient. If you are looking for a "best in class" list of fuel misers, go to www.ecoaction.gc.ca/vehicles. It has the winners, as well as a rating chart of every car on the market. When shopping for a vehicle, be sure to look at the EnerGuide label for that car's fuel efficiency.

65 Choose fuel-efficient options

Regardless of which new vehicle you choose, you need to consider your options carefully. Many of the fancy modern add-ons to cars are not good for gas consumption. To save fuel, here are five options to avoid:

1. **4WD/AWD (four-wheel drive/all-wheel drive).** This is probably the biggest drain on your car's engine. Turning all four wheels takes a lot of effort and fuel. Although 4WD or AWD may help with winter driving, would a good set of snow tires work just as well instead?

2. **V6/V8.** The more cylinders in your new car, the more gas it will burn up. Can you buy a four-cylinder engine for your everyday needs and rent a bigger-engined car for the annual summer camping trip?

3. **Air conditioning.** A car's air conditioning system burns a lot of fuel (see Tip 37). Can you survive without it? Can you opt for a sunroof or tinted windows instead?

4. **Big wheels and tires.** Although they look really cool, they are harder for your engine to turn—especially when made of steel. Opt for lightweight, aluminium alloy wheels to lighten the load.

5. **Power equipment.** You may be a bells-and-whistles kind of person, but the fewer options you choose, the lighter your car will be. Power seats, power windows, and other conveniences require energy to run and add weight to the car. Lighten up on the gizmos.

66 Buy a smaller car

As mentioned in Tip 32, excess weight has a huge effect on fuel economy. The more your car weighs, the harder your engine will have to work to move this excess mass.

> "One hard and fast rule of fuel economy is that heavier vehicles take more energy to move down the road and thus consume more fuel. Reducing weight will improve both fuel economy and vehicle performance."
>
> www.greencar.com/features/fuel-economy-fundamentals/

Small cars weigh less and have smaller engines than large cars. This saves a lot of fuel over the life of a car. Maybe it's time to switch to a smaller car. Do you really need a big V6 minivan when you have only two kids? Do you really need a V8 SUV just because you feel safer in it? Try to do more with less. Studies show that more than 70% of the time, a car has only one person in it—three to four seats are empty. Bigger is no longer better, especially with the price of gas these days.

67 Buy a 5-speed stick

In your search for a newer, more fuel-efficient car, keep in mind that another key element to savings is a manual shift. If you opt out of an automatic transmission, not only will you save on the price of the car but you'll save on fuel, too.

> "Among the small cars that we've tested, manual shift generally adds 2 mpg overall versus an automatic transmission."
>
> 2009 New Car Preview, Consumer Reports

A manual shift allows you to really control your car—great for the fun of driving, for safety, and for good gas use. With a stick, shift to a high gear as soon as possible to keep the rpms (engine speed) down. The less your engine is working, the less gas is used. Accelerate in a smooth fashion and keep in the highest gear possible without stalling, and be slow to gear down. (But don't rely solely on your brakes to reduce speed—this wastes fuel and will wear the brakes down quickly.) There will be no wear or strain on the engine, only great fuel savings.

68 Buy a hybrid

A hybrid vehicle combines two or more sources of power. This is not a new concept. Since the late 1800s, hybrid engines have been used in trains, ships, buses, and submarines. But for automobiles, it's a relatively new adaptation, one that combines a gas engine and an electric motor. When you drive a hybrid automobile at lower speeds (usually around town, up to 40 km/h), it will run on the electric engine only. Above these speeds, the electric engine gets an assist from the small gas-powered engine. When you take your foot off the gas pedal or come to a stop, the gas engine turns off again. The net benefit to this system is a tremendous savings in gas. For all slow speeds and crawling traffic, no gas is used. And the gas engine itself is a small, highly efficient one.

> "[Hybrids] offer greatly improved fuel economy (by as much as 40 to 50% compared to a conventional vehicle of the same size) and reduced greenhouse gas emissions."
>
> The Auto$mart Guide, Natural Resources Canada, 2007

Now that hybrids have become incredibly popular, prices have come down. That, combined with the assistance of government "green" cash incentives, makes hybrids less expensive than you might think. At current gas prices, the savings on fuel can offset the higher cost in less than three years. Many car manufacturers offer these modern fuel-sippers, and they come in all shapes and sizes. If a brand doesn't have a hybrid in its product mix, it is madly scrambling to come up with one.

A new and improved plug-in hybrid is coming in the very near future. This type of car will allow you to charge the battery overnight so that the first 100 kilometres driven the next day will be gas-free. At the end of the day, if any power is left in the hybrid's battery, you could also do the reverse and let your car power your home by plugging it in to your electrical supply. Don't believe all the negative rumours about hybrids; these cars have very few downsides. The hybrid has a critical role to play in solving the fuel crisis and global warming.

Here's a list of the top ten hybrids purchased in Canada over the past year (there was a tie for 7th place):

Model Name	Manufacturer	Litres per 100 km (city / highway)
Prius	Toyota	4.0 / 4.2
Insight	Honda	4.4 / 4.3
Civic (Hybrid)	Honda	4.7 / 4.3
Camry (Hybrid)	Toyota	5.7 / 5.7
Altima (Hybrid)	Nissan	5.7 / 5.9
Escape (Hybrid), 2WD	Ford	5.8 / 6.4
Malibu (Hybrid)	Chevrolet	7.9 / 5.8
Aura (Hybrid)	Saturn	7.9 / 5.8
Highlander (Hybrid)	Toyota	7.4 / 8.0
RX 400h	Lexus	7.8 / 8.4
Vue (Hybrid)	Saturn	8.2 / 6.1

MYTHS—AND TRUTHS—ABOUT HYBRIDS

There are many naysayers out there who want to bash the hybrid movement. These pessimists want to discredit hybrid vehicles for a lack of power or for poor battery life or for the cost, but they are ill-informed. Here are some of the rumours—and truths—about these wonderful fuel-saving machines:

Myth: Hybrids are expensive.
Truth: Hybrids do cost more than a conventional car, but there are government tax credits and incredible fuel savings in return. The more expensive gas gets, the quicker the extra cost of a hybrid is paid off. The Honda Insight and the Toyota Prius (the most fuel-efficient cars on the road today) aren't too expensive, both being under $25,000.

Myth: Hybrids have poor fuel economy on the highway.
Truth: The hybrid system is most effective for city driving. This is where the electric engine takes over and the gas engine is off. When you increase speed or drive on the highway, the gas engine comes on. When this happens, fuel economy goes down, but it's still not bad. A hybrid is amazing in the city and good on the highway. In fact, hybrids are still better on gas on the highway than most gas-powered cars. In the end, the net fuel consumption is what counts, and hybrids are much better than conventional cars for both city and highway driving combined.

Myth: If you crash a hybrid, there is risk of fire or electrocution.
Truth: There's a risk with any car when it crashes, particularly when there's a tank of gas on board. But in a hybrid, the electric components and the battery pack are

completely sealed and protected from rupture. These systems have been thoroughly tested by the manufacturers and government motor vehicle agencies. These cars would not be allowed on the road if there were any potential risk to the occupants of the car.

Myth: When the hybrid battery runs out, you will have to scramble to plug it in and recharge it.
Truth: The hybrid battery is regenerative, meaning it recharges itself as it runs. The coasting motion of the car's spinning wheels and the excess brake energy that's created when you stop the car is harnessed and sent back to the battery to replenish its power. If for some reason the battery does get drained, the gas engine takes over and powers the car, which will then recharge the battery.

Myth: Hybrids have no power on the highway.
Truth: A hybrid vehicle actually has two (or more) engines on board. When you are driving a hybrid on the highway and ask your engine for a bit of *oomph* to pass another vehicle, you have a gas engine and an incredibly powerful electric engine working together. According to Toyota, the Lexus RX 400h hybrid has 270 hp on tap. This is more than adequate for any highway speed or passing power.

Myth: When a hybrid battery dies, it goes into landfill.
Truth: With modern technologies, these batteries can be mostly recycled. The materials inside are not as toxic as some would like you to believe. But even if the batteries couldn't be recycled, given that most hybrids use 40% less fuel than other cars and emit 40% less harmful CO_2, the leftover battery is a very small price to pay for the massive reduction in fuel usage and pollution that these cars offer over the years.

69 Buy a new diesel

Diesel-engine cars have changed dramatically since the smelly, smoky, and noisy models of the 1970s. Because of many technological innovations in engine development and a reduction in sulphur in diesel fuel, these cars are now called "clean diesels." With many of these modern diesels, it's difficult to tell if they are gas- or diesel-powered when they are running. There is virtually no smell, no smoke, and little sound. Diesel cars offer incredible fuel savings.

> "Diesels get about 30% better fuel-economy than gasoline powered cars and have lots more pulling power."
>
> 2009 New Car Preview, Consumer Reports

Diesel cars can go much farther on a tank of fuel than gas-powered cars because diesel contains much more energy than regular gas. As well, diesel is efficiently burned in the engine by very high compression instead of by a spark plug, so you get much more power, performance, and distance out of it. Diesel vehicles usually cost more than a regular gas-powered car, and diesel fuel costs more than regular fuel. However, much like the hybrid cars, you'll see the payoff from increased mileage in only a few years.

70 Buy an old diesel

If you can't afford a new diesel but still want to switch to a more fuel-efficient vehicle, why not buy an older-model diesel? Although these earlier automobiles may be a bit smoky, have the odd rattle, and give off an oily smell, they still get excellent gas mileage.

> "The diesel engine is the most efficient of all the types of internal-combustion engines, meaning that it extracts more mechanical energy from its fuel ... it is the increased energy that gives diesel-powered vehicles their edge in fuel economy."
>
> www.edmunds.com/advice/alternativefuels/
> articles/102946/article.html

The older diesel cars can also be converted to run on biodiesel. This new fuel is made from soybean oil, animal fat, or used restaurant grease. A converted diesel is truly clean, as there are very few fumes emitted from its tailpipe. The car's exhaust can smell like french fries! The best part is that the fuel for these cars can be free. Most chip stands and restaurants are glad to give you their used vegetable oil. Gas stations are now starting to offer this innovative new fuel in its pure form or mixed with regular diesel fuel for added performance.

71 Buy a flex-fuel ethanol car

Ethanol is an alternative biofuel that can be produced from corn, sugar cane, and forestry and agricultural waste. It has been around for a long time; in Brazil it's been used for years. But it's all very new to North America and is becoming especially popular in the American Midwest, where corn is grown. Many of today's modern cars can run on E85, an 85% ethanol, 15% gasoline blend. These cars have a badge on the back that reads "flex fuel." This means that their engines are flexible and can run on ethanol or normal gasoline. Unlike hybrid and diesel engines, you don't have to pay extra for this type of car. Ethanol is also much cheaper than regular gas. In many places in the United States, ethanol is on average 18% cheaper than regular gasoline. (It is still hard to come by in Canada.) However, there are a few major drawbacks to ethanol:

Cars running on E85 get 27% worse fuel economy than gas cars.

Less than 1% of U.S. gas stations sell E85.

Increasing demand for ethanol has been blamed for driving up food prices.

2009 New Car Preview, Consumer Reports

Although the emissions are lower and the fuel is cheaper, ethanol may not be the right answer to fuel savings.

72 Convert your car to CNG or LPG

CNG (compressed natural gas) and LPG (liquid propane gas), better known simply as natural gas and propane, are two fuels that have been around for decades. They are quite similar and are the most common alternative fuels used today. Cheaper than gas, they also have less of an impact on the environment.

> "The average price of natural gas can be as much as 40% cheaper than gasoline.... Propane is on average 25 to 35% less expensive than gasoline in Canada."
>
> The Auto$mart Guide, Natural Resources Canada, 2007

To use CNG or LPG in a car, you can either buy a new model that runs on these fuels or you can convert your existing vehicle with a kit. Find a mechanic to install the kit—it's quite technical. The conversion can cost between $2,000 and $4,000, but there may be government assistance to help lessen the burden. Regardless, the conversion cost is easily recouped through reduced fuel costs. Unlike ethanol, the fuel consumption is the same as with a gas-powered car, the fuel is readily available, and using it won't make the price of corn go through the roof.

PART 10

Advanced
gas-saving
tips

73 Bonus tips

There are so many more tips to save your fuel than the 72 I've listed so far. But there's room for only 75 in this compact book. So in this tip I've gathered several ideas about how to keep money in your pocket:

- **Drive in the slipstream.** When driving on the highway, don't tailgate the car in front of you but instead get behind a big truck or row of cars that are cutting through the wind at a constant speed. Let them do the work and ride their slipstream (also known as "drafting") to give your engine a break and save fuel.
- **Anticipate green lights.** Most stoplights are computerized to change in a sequence. At a green light, drive easily and at the speed limit, maintaining your speed smoothly. Most likely, you'll get all the greens working for you. If not, slow down a bit to get into the sequence. Racing up to the next light does nothing but waste gas.
- **Build in a buffer.** When driving on the highway, drive in the slipstream but also leave a big gap (two car lengths) between you and the cars around you. This gives you the option of easing off the gas and back onto it with the ebb and flow of traffic. If you are jammed in with many other cars, you'll be stopping and starting, and burning up lots of unnecessary fuel.
- **Drive properly on hills.** Never "floor it" on a hill to gain speed. This wastes fuel. Start slowly, building speed before you get to a hill, maintain the

speed on the hill, and then coast down the backside of the hill with your foot off the gas pedal. According to TreeHugger, a sustainability website, you can save up to 30% of your fuel by doing this.

- **Get a locking gas cap.** These caps are easy to find and simple to use. With gas so expensive these days, thieves stealing fuel from gas tanks is becoming a big problem in many areas. Lock it up and rest easier.

- **Buy a light-coloured car.** We all know that a darker coloured car attracts the sun and the heat. (It also seems to attract more scratches and looks dirtier.) So, for your next car, buy a light-coloured car to help keep the inside cool. It will also help keep the gas tank cool, so you won't lose as much gas through evaporation.

- **Avoid dirt or gravel roads.** When you travel on unpaved roads, your car has to work harder to compensate for the spinning of the tires. These roads are slippery and you waste fuel as the engine runs faster. Also avoid driving on snowy or really rainy days—lots of slipping and sliding burns up fuel.

- **Listen to calm music.** Listen to peaceful, relaxing music instead of wild or loud stuff. The crazier the music in your car, the crazier you'll drive. Chill out and be calm, your pocketbook (and other drivers) will thank you.

74 Try hypermiling

Do you think that you can master the previous 73 tips and save hundreds of dollars a year in fuel? Well, here's a way to really challenge yourself: Try "hypermiling." Hypermiling is a new term for extreme ways to save fuel. Check out www.cleanmpg.com to see how Wayne Gerdes is revolutionizing the way we drive our cars. In his quest for the ultimate in fuel economy, he slipstreams his car behind big trucks, turns his car off to roll down hills, pushes his car out of the driveway—anything to save a drop of precious, expensive fuel.

> "Seems like an awful lot of trouble to save a little fuel. But Gerdes saves more than a little. He says he's managed to get 59 mpg out of his regular 2005 Honda Accord, which is rated at 34 mpg. Put Gerdes in a Prius, and he once recorded 127 mpg."
>
> www.marketplace.publicradio.org/display/web/2007/04/27/hypermiling

Some people think Gerdes is crazy, others think he is dangerous and breaking the law. No matter what the case may be, he and hundreds of other CleanMPG bloggers are continually coming up with creative ways to save fuel.

75 Start a fuel-economy group

Now that you have read this book and become an expert on saving fuel, it's time to spread this wonderful knowledge with your friends, family, neighbours, and colleagues. Why not form a fuel-economy group? You don't need to be as radical as the hypermiling folks, but you could organize others to share creative ideas and tips on how to save fuel. EcoModder.com is an established fuel-economy group similar to CleanMPG. Members can log on to the group's site to discuss new angles and ways to obtain better fuel economy.

> "EcoModder.com is an automotive community where performance is judged by fuel economy rather than power and speed. EcoModders employ a combination of vehicle modifications, driving techniques, and common sense to squeeze every penny out of the pumps."
>
> www.ecomodder.com

Plan to meet monthly over coffee, start a blog together, or link up in a group email. Whatever the medium, challenge one another and make it a game to see who can get the best mileage. Have a trophy or gas card reward for each month's best "fuel saver" member.

PART 11

Internet
resources

Internet resources

There is an abundance of resources on the internet that can help you save fuel and money, from automobile clubs to environmental sites. Here's a selection to get you started.

www.aaa.com: Similar to the Canadian Automobile Association, but based in the United States.

www.autoshare.com: This Toronto-based auto-sharing company has been around for more than 10 years. As pioneers in this relatively new industry, its business success is based on providing 24-hour self-serve access to cars and vans.

www.autotrader.ca/www.autotrader.com: When searching for a more fuel-efficient used car, the online *AutoTrader* is a great place to start. It has used-car classified ads and pricing, and links to "all things automotive."

www.caa.ca: The Canadian Automobile Association offers emergency roadside assistance, insurance, and travel maps to members, among other benefits. It also advocates for issues related to automobiles and motorists.

www.carheaven.ca: If you have an old, fuel-inefficient car that needs to be disposed of, this is the website for you. CarHeaven is a charitable Canadian program designed to remove these high-polluting cars from our roads.

www.carproof.com: When looking to buy a more fuel-efficient used car, you may want to consult this website, which provides history reports on many cars in Canada, to make sure the one you have in mind hasn't been damaged or stolen.

www.cleanmpg.com: For diehard fuel economists. This site is an open discussion forum "dedicated to increasing fuel economy, mileage (MPG), and lower emissions."

www.consumerenergycenter.org: The California Energy Commission website gives up-to-date information on energy resources and tips on how to use them wisely at home, at work, and in our vehicles.

www.consumerreports.org: This unbiased organization rates and reviews every vehicle on the North American market, providing expert consumer advice and recommendations. It also has a massive membership; members submit their own surveys of vehicle quality.

www.dailyfueleconomytip.com: This website is the result of one man's quest to provide his viewers with a fuel-saving tip each and every day. His goal is to "pay attention to easy ways we can save fuel."

www.davidsuzuki.org: David Suzuki's Canadian foundation website has excellent ideas for environmental conservation. It strives to "protect the balance of nature and our quality of life."

www.eartheasy.com: Another resource for sustainable living. Eartheasy's goal is to show us "how to simplify our life by living lightly on the earth."

www.ecomodder.com: This website forum is all about saving fuel. It offers "fuel economy modifications, hypermiling, and ecodriving for better MPG."

www.edmunds.com: Another site to visit when you're considering changing to a newer, more fuel-efficient vehicle. It offers car comparisons, purchasing advice, and car reviews and ratings.

www.epa.gov: The U.S. Environmental Protection Agency's website has many ideas on conservation and greening our world. The agency's mission is to "protect human health and the environment."

www.gasbuddy.com: A great way to find out about "local, real-time gasoline prices." Plan ahead for your next fill-up by using this website to find which stations in your area have the cheapest gas.

www.greenercars.org: When looking for a newer, greener car, consult this website. It provides news and consumer information to help you choose a vehicle that does "the least harm to the planet."

www.mapquest.com: This website offers maps and driving directions to help you plan your trip. It also provides links to hotels, restaurants, and businesses near your destination.

www.maps.google.com: Another online map source for planning trips.

www.oee.nrcan.gc.ca: Natural Resources Canada's Office of Energy Efficiency website offers tips on saving energy (and money) while helping to protect the environment.

www.pickuppal.com: This worldwide commuting website helps both drivers and passengers connect online.

www.smartcommute.ca: This Toronto-based commuting program is offered by Metrolinx, an agency of the Ontario government. It's a great example of how business, government, and the community can work together to fight traffic congestion.

www.tirerack.com: This is a great resource when you are shopping for new all-season or snow tires. It has a big inventory of discount tires and wheels from which to choose and provides reviews and ratings of tires.

www.treehugger.com: This website offers numerous ideas and thoughts on the environment and savings. Its focus is to "share sustainable design, green news, and solutions."

www.zipcar.com: The world's largest auto-sharing company, it operates out of many major cities in Canada and the United States.